CELL WARS

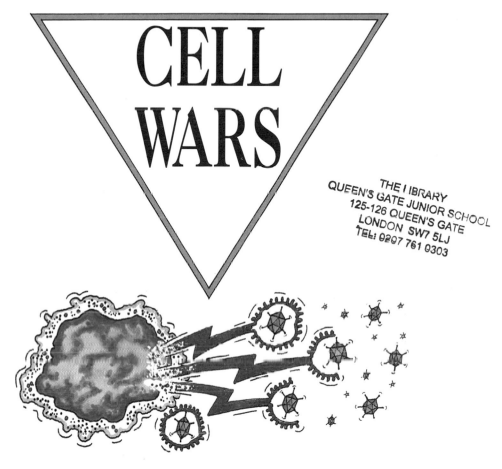

written by
Dr Fran Balkwill
illustrated by
Mic Rolph

Collins

William Collins & Sons Co Ltd
London • Glasgow • Sydney • Auckland
Toronto • Johannesburg

First published 1990
©text Fran Balkwill 1990
© illustrations Mic Rolph 1990

A CIP catalogue record for this book is available from the British Library

ISBN 0 00 191164 – 3
ISBN 0 00 196307 4 (PB)

Printed and bound in Portugal by Resopal
This book is set in Lubalin Graph 13/16

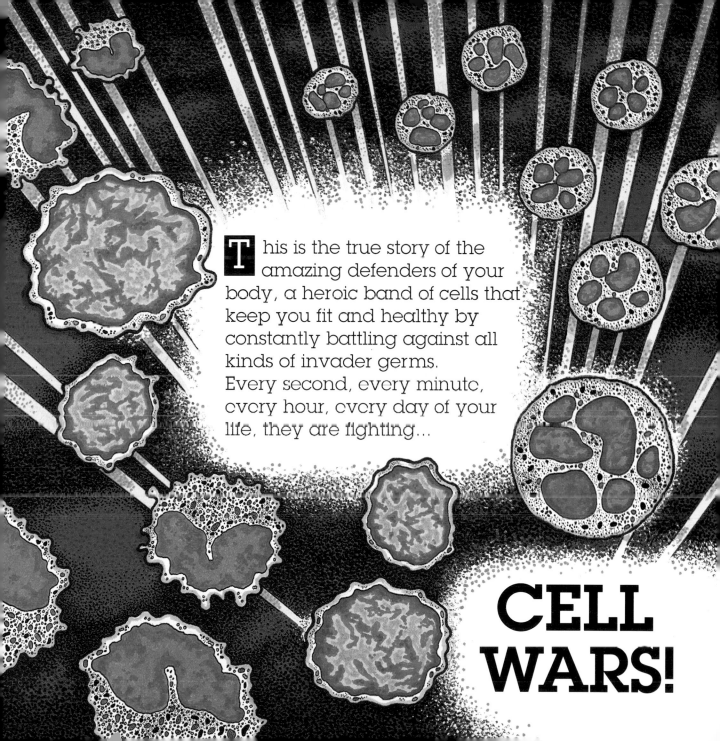

This is the true story of the amazing defenders of your body, a heroic band of cells that keep you fit and healthy by constantly battling against all kinds of invader germs.
Every second, every minute, every hour, every day of your life, they are fighting...

CELL WARS!

Every part of your body is made of tiny cells – about one hundred million million of them! You have muscle cells that make you move, nerve cells that make you see, touch, hear, smell and taste things, and bone cells that stop you going all floppy. Cells make your hair and your teeth. Cells make your nose and your eyes.

Skin cells are very clever. They make a tough, dry layer of dead cells all over your body which protects you from bumps, thumps and germs.

But sometimes crafty germs find a way into your body, through your mouth, nose, or ears, or even through a cut in your skin. That's when you need your defender cells!

Every day of your life about one thousand million defender cells are made in the marrow in the centre of your bones.

Can you believe that ?

MY HEROES!!

4

THE DEFENDERS

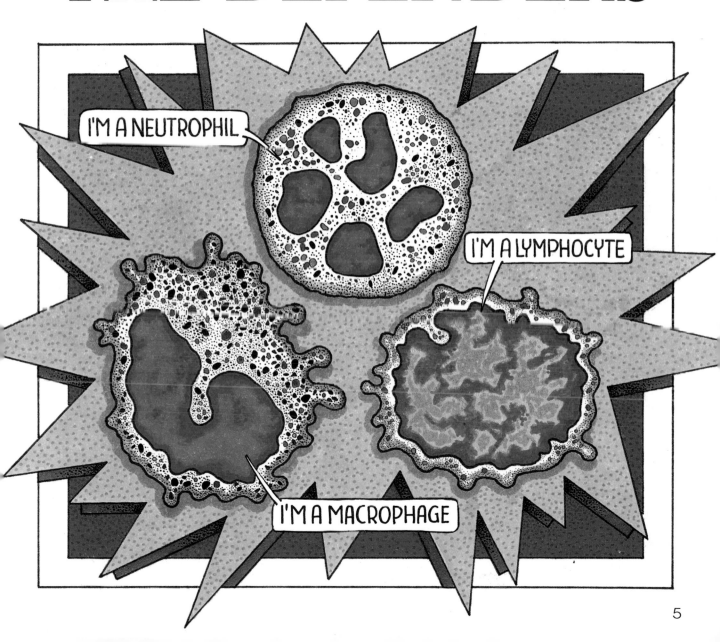

Neutrophils (New-tro-fils) are cells that are full of germ destroying chemicals. They travel round the blood stream armed and ready to destroy germs that can make you ill.

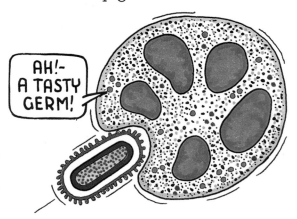

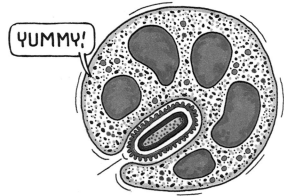

When a neutrophil detects an invading germ...

it gobbles it up and zaps it with its deadly chemicals.

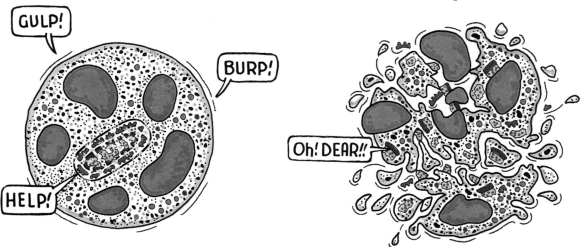

This destroys the germ but, unfortunately. . .

the neutrophil is sometimes destroyed as well.

Don't worry though, there are plenty more where that one came from !

Macrophages (Mac-ro-fay-jes) are mobile rubbish disposal units – they clear up whenever you are ill or injured, or wherever germs and dirt collect.

They travel around your blood stream and then settle down in many important parts of your body. There are millions in your lungs, for instance – continually eating up dust and germs that you breathe in every day.

Like neutrophils, they are full of germ zapping chemicals, but they live a lot longer. Macrophages don't usually self destruct.

Before they completely destroy the germ they have captured, they show it to the other defender cells – the lymphocytes (lim-fo-sites). This warns the lymphocytes that there's trouble afoot, and they join the battle.

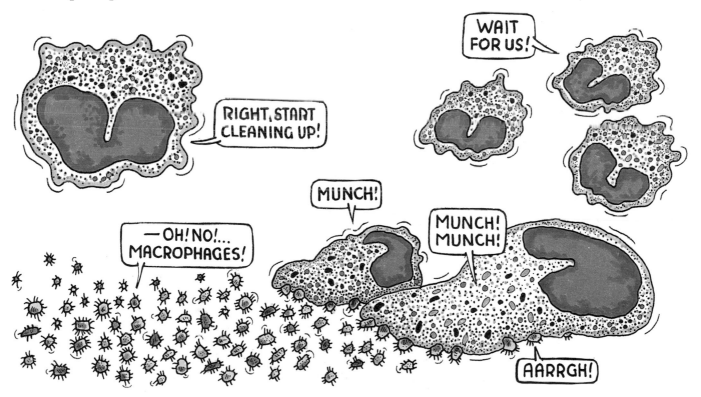

Lymphocytes are the cleverest of all your defender cells. They don't all attack every germ they meet. Instead, different squads of lymphocytes attack different germs.

Every type of germ that invades your body carries its own identification marks, and you have a lymphocyte squad already programmed to fight each of these. Amazingly, these lymphocyte squads were formed in your body before you were even born!

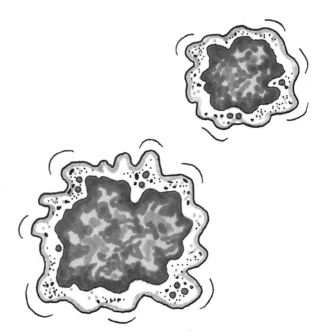

A lymphocyte squad first meets the germ it has been programmed to fight, and quickly multiplies into a powerful army.

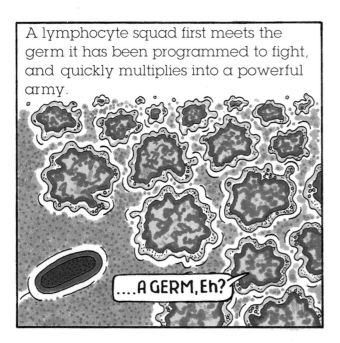

....A GERM, Eh?

Some cells in the lymphocyte army release special weapons called antibodies.
Antibodies stick tightly to the germs and stop them doing more harm.

OOOF!

Other cells in the lymphocyte army turn into destroyers. Their task is to seek out and kill off cells that have been invaded by germs.

The struggle is over and the lymphocyte squad is stronger than before. Some of these lymphocytes become memory lymphocytes and remain on alert deep in the body for years.

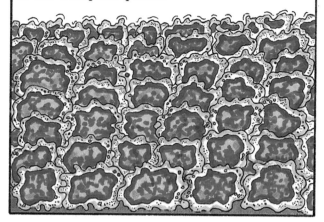

If the same germ tries to invade again, the lymphocyte squad is much stronger. So the battle is won more easily. Which explains why most children only have one attack of measles or mumps.

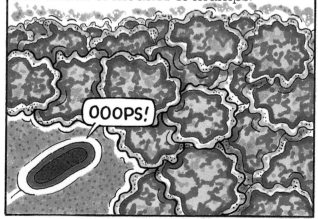

Neutrophils, macrophages and lymphocytes must all fight side by side to win **Cell Wars.**

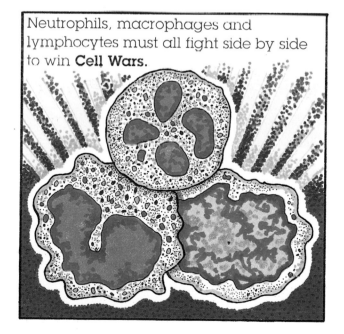

Viruses are public enemy NUMBER ONE – unseen alien invaders of your body. They come in some weird and wonderful shapes

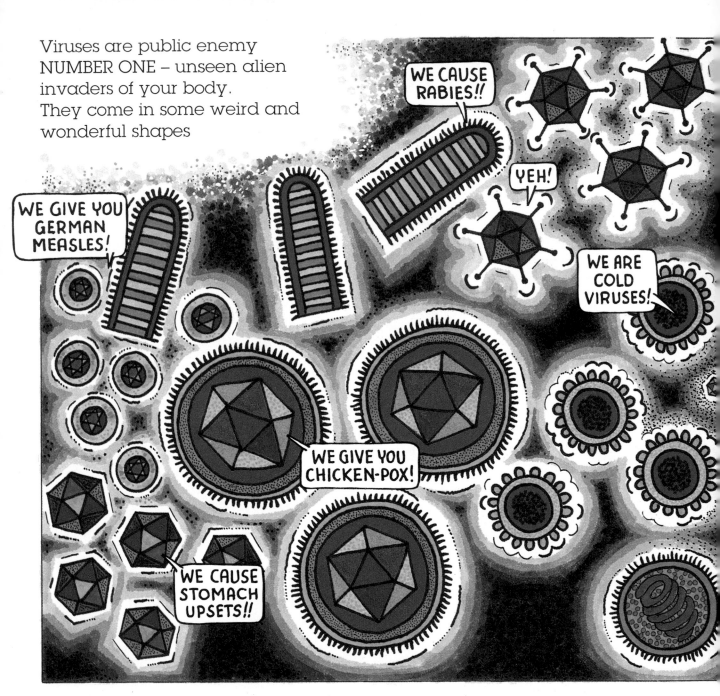

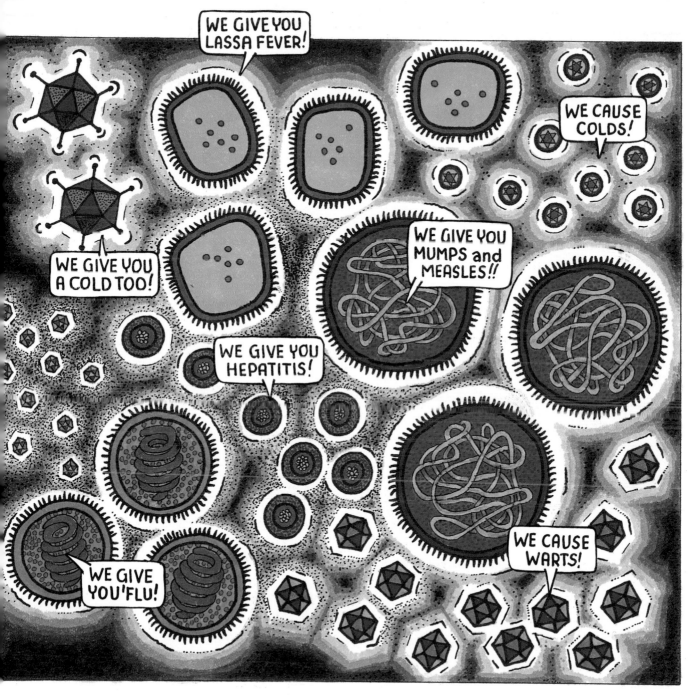

Viruses are germs that make you ill by entering your body, invading some of your healthy cells, and turning them into virus making cells.

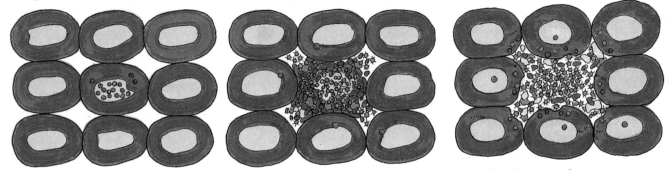

Instead of doing their proper jobs, healthy cells are made to produce viruses that invade more and more cells,

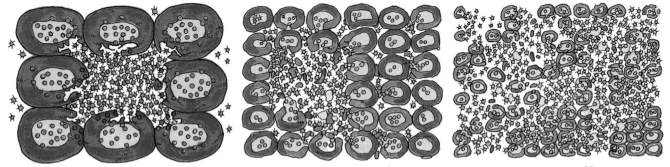

that make more and more viruses that invade more and more cells . . .

Viruses are very, very, small, much smaller than the cells in your body. If one of your cells was this big (1cm) you'd be as tall as the Empire State building in America, but if a cold virus was this big (1cm). you'd be sneezing amongst the stars !

BLESS YOU!

13

WATCH OUT ! !
Millions of cold viruses are now
on their way to invade and
destroy the cells of your nose and
breathing tubes.

COUGHS and SNEEZES
SPREAD DISEASES!

Let's look up your nose
and see what happens . . .

14

The cells inside your nose are always prepared for trouble. They make a gloopy fluid called mucus that viruses and other intruders get stuck in. What's more – mucus contains those special lymphocyte weapons, antibodies. Tiny hairy bits on the nose cell surface waft the germs, antibodies and mucus away.

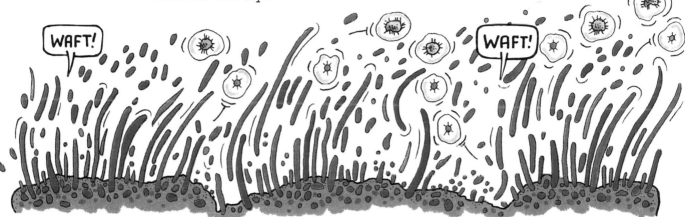

But sometimes a few viruses sneak through the mucus and invade the cells inside your nose. They turn the nose cells into virus making cells. Lots more viruses then attack lots more cells. Some nose cells die and others produce watery stuff that makes you sneeze and feel all bunged up.

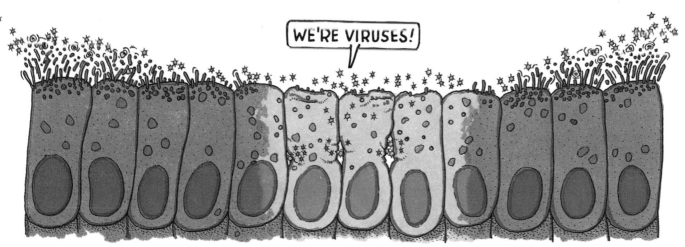

Don't worry! Those viruses are soon spotted by your defender cells. They rush from the tiny blood vessels in your nose and try to destroy as many viruses as they can.

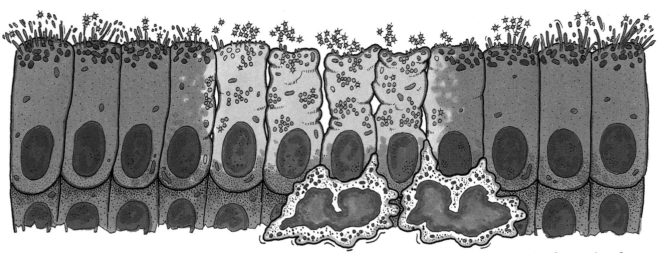

Macrophages, neutrophils, as well as the nose cells, send out chemical alarm signals. These signals protect nearby cells from virus invasion and summon lymphocytes to the trouble zone.

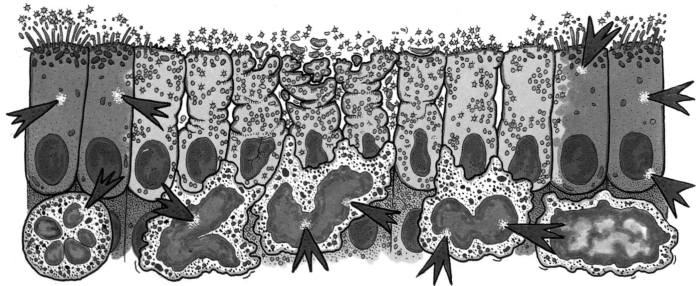

Now on full alert, the cold virus squad of your lymphocytes is rapidly multiplying. Soon you have thousands and thousands of lymphocytes programmed to destroy virus occupied cells.

Lymphocytes release antibodies that coat the cold virus. Once a cold virus is coated with its antibody it can't invade any more cells, and, what's more, it gets gobbled up in double quick time by neutrophils and macrophages.

So it's not surprising that colds and other virus infections get better, when you've got all those hard working defender cells inside you.

But your defender cells have other enemies . . . bacteria !

Bacteria are public enemy NUMBER TWO – They are smaller than human cells but much bigger than viruses. You find bacteria everywhere – in the air, in the sea, in rivers, in the soil, on your skin, on your teeth, even inside you! Most of these bacteria are not your enemies. In fact, they are an important part of nature.

Bacteria that live in the soil, for instance, break down dead plants and animals.

But a few types of bacteria are very harmful. If they get inside your body they multiply rapidly, releasing poisons that can make you feel ill.

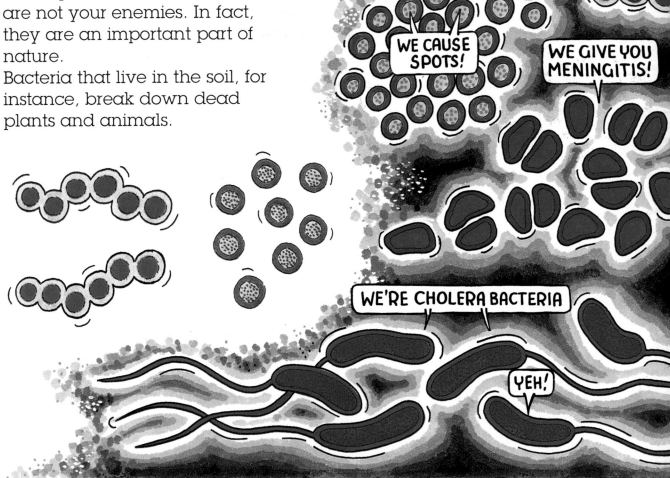

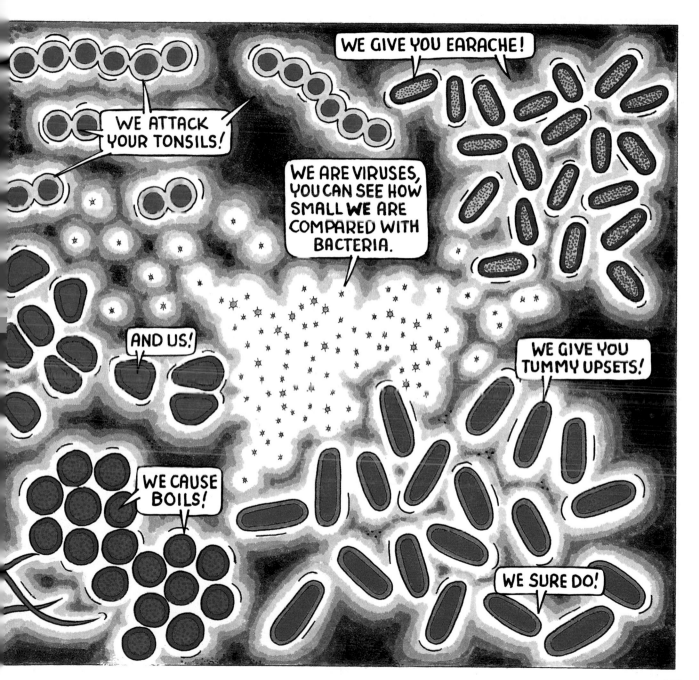

19

One way that harmful bacteria can get inside you is in food that has not been stored or cooked properly. Imagine this...

After a Saturday shopping trip you put a chicken and some ham in the fridge. A few dangerous bacteria are lurking inside the raw chicken, but they can't multiply in the cold fridge. The chicken is slowly dripping onto the ham. The ham now has some of those sneaky bacteria on it – not enough to harm you, that would take at least a million.

DRIP! DRIP!!

On Sunday you make some ham sandwiches for school lunch next day. Overnight in the kitchen the bacteria start to multiply.

After twenty minutes each bacteria has become two, after another twenty minutes, two become four.
By the morning there are over two million sneaky bacteria in your ham sandwiches!
You enjoy your ham sandwiches at school – little do you know what else is inside them! During the afternoon the bacteria keep on multiplying inside you. They stick to the cells that line your stomach and make poisons that destroy your cells.

GREAT!!

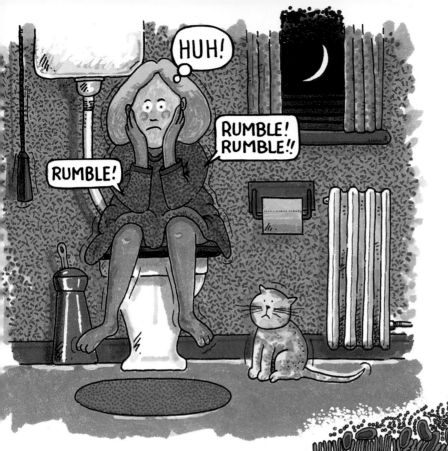

In the middle of the night you feel really ill!
Don't panic ! Your defender cells fight back. Neutrophils and macrophages find bacteria just as tasty as viruses.

And lymphocytes love a good fight with bacteria.

Although you are feeling groggy, your defender cells are still fighting the bacteria and their poisons. With some rest and lots to drink you soon feel much better.

SO REMEMBER
Keep food covered in the fridge. Raw foods may contain some dangerous bacteria which are killed by proper cooking.
Wash your hands, the kitchen tools, and the kitchen surfaces, after handling raw food.

Scientists and doctors have found a really clever way of protecting you from viruses and bacteria, with help from your defender cells.
It is called vaccination (vax-in-ation).
They have created harmless dummy germs in the laboratory.
These dummy germs look just like the real germs
that doctors want to protect you from.

When the dummy germs are given to young children, a lymphocyte squad is immediately alerted. This quickly multiplies into a strong army, that attacks the harmless dummy germ. So if the real germs ever attack you after you have been vaccinated, you already have an army of lymphocytes to protect you. Unfortunately scientists haven't been able to find vaccines (vax-eens) for all troublesome germs. The crafty 'flu virus keeps on changing its identification marks and there are over a hundred different viruses that cause colds!

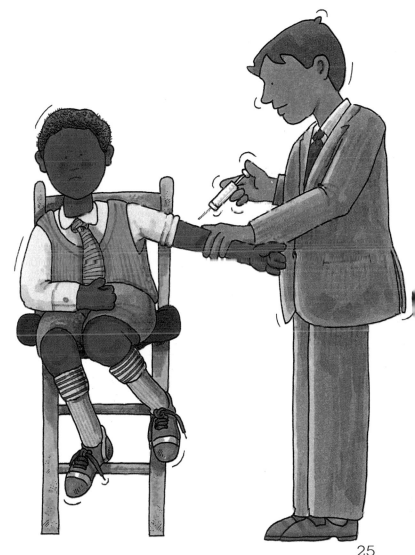

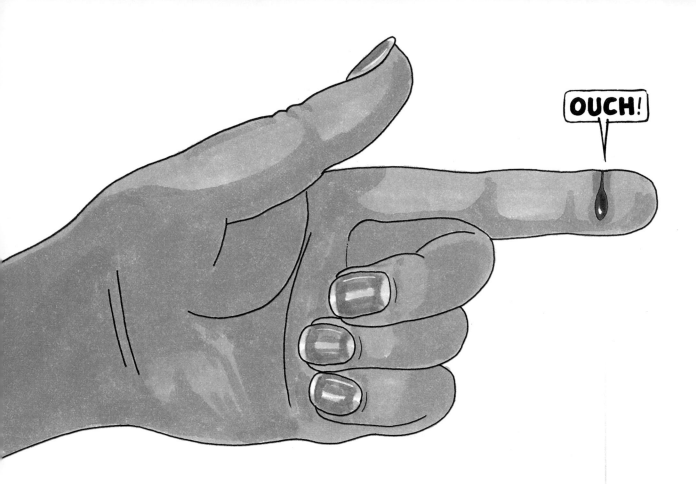

Defender cells don't just work together to fight illnesses, they also help repair your body if it gets damaged.
Imagine you've just cut your finger. It bleeds because you have cut through some of the tiny capillaries that carry blood amongst your skin cells.
The blood soon becomes dark and sticky and stops flowing, forming a blood clot. The blood clot dries and shrinks, pulling the edges of the cut together – that's what a scab is.

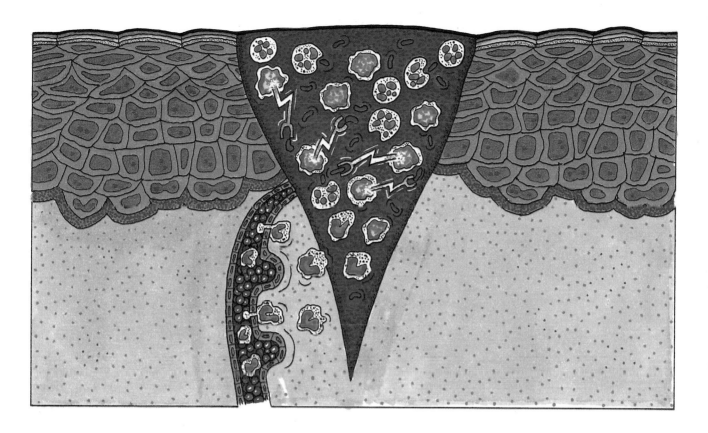

If harmful bacteria and viruses get into the cut, they must be zapped! Chemical messages released by injured cells alert neutrophils and lymphocytes. They leave the blood stream by the thousand and move into the blood clot, hunting for germs.

After twenty-four hours the macrophages start to move in. They get to work, cleaning up dead cells, live bacteria, and clotted blood. Blood vessel cells begin to divide and make a new blood supply for the cut.

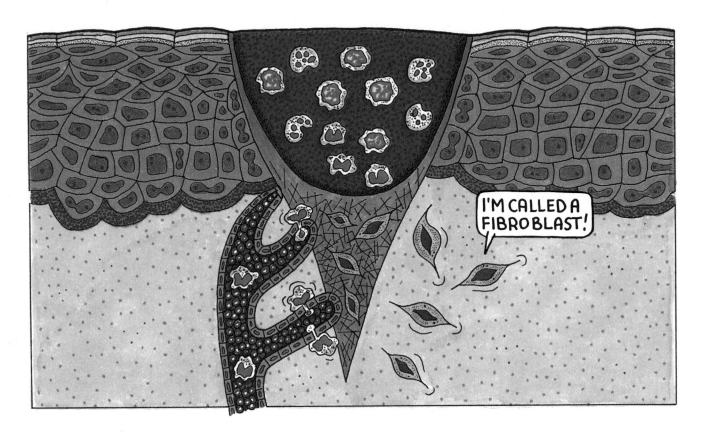

After about three days, your defender cells have cleaned up the cut really well and all danger of infection is past. It is then that the builder cells, fibroblasts, are brought into action. Fibroblasts are long thin cells that make special strengthening strands that pull the broken skin firmly together.

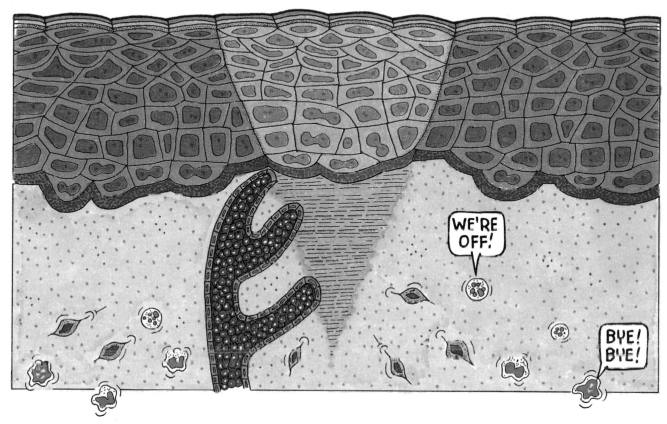

Slowly the defender cells and fibroblasts disappear. They have done their job. Now the skin cells take over, growing out from the edge of the cut.

Sometimes, however, a wound might be deep and your cells cannot cope so well. Then a doctor can help your body repair itself by putting in some stitches.

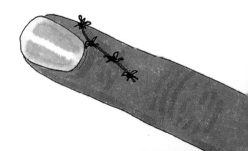

This isn't the whole story of your defender cells – they also fight bigger enemies like funguses and tapeworms, they help broken bones to mend, they eat up dead and dying cells and make spots vanish. Now you don't have to worry about invader germs that try to make you ill! You know that your defender cells are always ready to fight **CELL WARS**!!

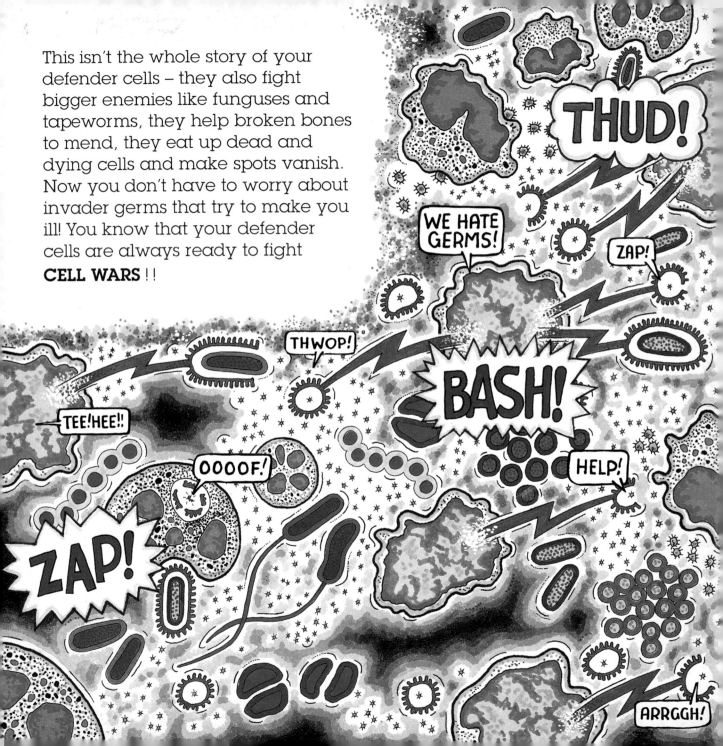

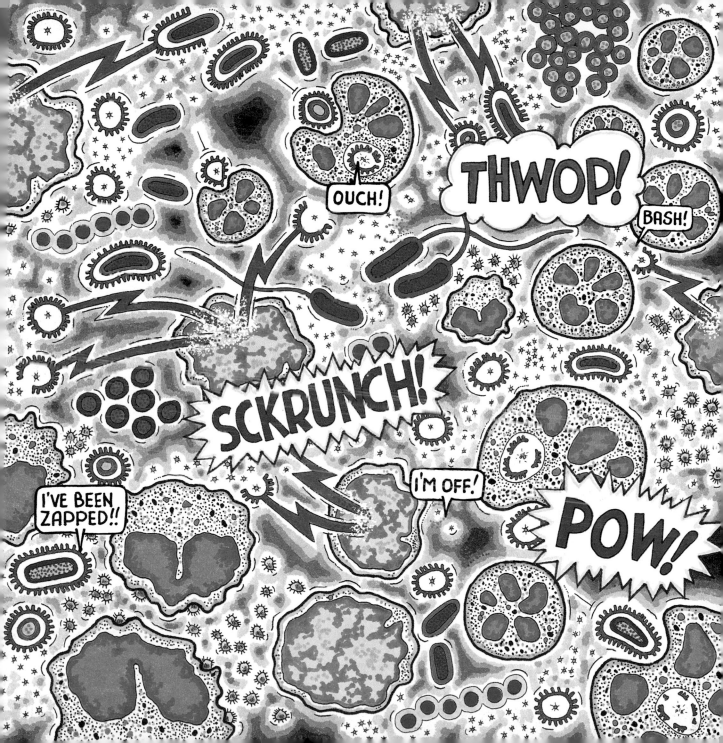

THE DEFENDERS

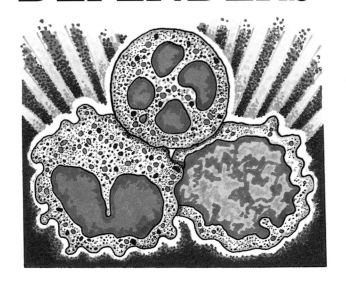